Environmental Division 2021

Held at the 2021 AIChE Spring Meeting and 17th Global Congress on Process Safety

Online
18 – 23 April 2021

ISBN: 978-1-7138-5339-8

Printed from e-media with permission by:

Curran Associates, Inc.
57 Morehouse Lane
Red Hook, NY 12571

CURRAN ASSOCIATES INC. proceedings.com

Some format issues inherent in the e-media version may also appear in this print version.

Printed with permission by Curran Associates, Inc. (2022)

AIChE
120 Wall Street, FL 23
New York, NY 10005-4020

Phone: (800) 242-4363
Fax: (203) 775-5177

www.aiche.org

Additional copies of this publication are available from:

Curran Associates, Inc.
57 Morehouse Lane
Red Hook, NY 12571 USA
Phone: 845-758-0400
Fax: 845-758-2633
Email: curran@proceedings.com
Web: www.proceedings.com

TABLE OF CONTENTS

617749 Evaluation of the Expanded Polystyrene Pyrolysis Process

Abstract:

Expanded polystyrene (EPS) is a material in great demand in the industrial, food and construction markets, among others. Due to this, it is one of the most abundant and invasive waste for the planet, therefore, it is necessary to search for measures that help in the mitigation of said waste. For Colombia, the EPS discarded is part of 13% of the plastics discarded in landfills (3,400 tons / day or 102,000 tons / month approximately). "Colombia is among the 190 countries where the amount of EPS disposed of in landfills is not controlled, an amount that increases according to population density. From this material, important precursors can be obtained by thermochemical transformations. In this context, the present research had the general objective of evaluating the obtaining of fuel precursors from the pyrolysis of EPS. The raw material was obtained from household appliance packaging and food packaging. The material underwent the drying, grinding and sieving process. Subsequently, it was characterized physicochemically with the techniques of proximate analysis, Fourier transform infrared spectroscopy (FTIR), thermogravimetric analysis (TGA), ultimate analysis and calorific value. From the foregoing, the ideal conditions were established to carry out the pyrolysis. The pyrolysis time was one (1) hour and the product that was obtained the most was solid, followed by gas and liquid, with special interest in liquid product, these products were characterized by calorific value, FITR and elemental analysis to the solid product.

617845 Adsorption of Ibuprofen Using Activated Carbon Obtained from Tire Waste

Abstract:

The management of urban solid waste offers possibilities to implement environmental alternatives that integrate the protection of hydric resources and serves as raw material to generate solutions in the energy industry.

This study explores how to get activated carbon from used tires, becoming a viable alternative since only in Bogotá 3 million tires are thrown away per year, which constitutes a vector of rodents, insects, fire risks, and a disease focus. Two types of activated carbon from the used tires were prepared, one through physical activation with nitrogen (N2), and another by chemical activation with phosphoric acid (H3PO4) using a 3:1 ratio (active agent/carbon) in both of them. The absorbent was tested in the removal of a probe molecule belonging to the CCL4 list.

The raw material was characterized by thermogravimetric analysis (TGA), proximal and elemental analysis, and infrared analysis with Fourier transform (FTIR). For the obtained activated carbon textural characterization was made to determine surface area, volume and pore distribution, the isoelectric point was quantified also.
These carbons were used to remove ibuprofen from the aqueous solutions of 20, 50 and 80 mg/L by varying the temperature and thus establishing the absorption capacity of the carbons for this contaminant.
The results demonstrate that the chemically activated carbon has a higher superficial area than the obtained by physical activation, the amount of pollutant removed is determined for gas chromatography with a mass spectrometer.

Adsorption Kinetics of N-nitrosodimethylamine (NDMA) onto Activated Carbons

Kailey Garland, David Amidei, Paul A. Andersen, Catherine E. Brewer

According to the U.S. Environmental Protection Agency, N-Nitrosodimethylamine (NDMA) is classified as a probable human carcinogen. A risk assessment indicated that a concentration of 0.7 ng/L (parts per trillion) of NDMA in drinking water represents a 1 x 10^{-6} increased cancer risk level. NDMA is highly mobile in soil, and was derived from the treatment of water from rocket fuel testing, which resulted in the contamination of ground water near rocket testing sites. Additionally, NDMA can be an unintended byproduct of the chlorination of wastewater, making it a significant concern for wastewater treatment plants that use chloramines for disinfection. The treatment standards for NDMA typically range from 1 to a few ng/L, which creates analytical challenges for monitoring and development of remediation methods.

The purpose of this research was to test the ability of biochars made from local biomass wastes to adsorb NDMA from water at treatment-relevant concentrations. The adsorption kinetics of biochars produced from pecan shells and local forestry residues were tested and compared with a commercially available coconut shell-derived activated carbon. Further testing of the commercially available carbon was conducted in a lab-scale column for adsorption of NDMA to be used as a basis for future work. Concentrations of ^{14}C-labeled NDMA in well water from the NASA White Sands Test Facility (Las Cruces, NM) were measured at regular time intervals over the course of the flow through a biochar-filled column. Liquid scintillation counting was used for NDMA quantification.

Freundlich isotherm models were used to estimate column dimensions, flowrate, and breakthrough times. The resulting column was 18 inches (45.72 cm) long with a diameter of 0.5 inch (1.27 cm), and used an initial flowrate of 240 mL/min for a breakthrough time of 6 hours. Experiments showed breakthrough occurring immediately, however, despite changing experiment parameters such as flowrate and carbon preparation.

Calgon coconut shell carbon (CSC) provided the fastest and highest capacity for NDMA adsorption. Synthesized pine bark char adsorbed more NDMA by the end of the kinetics study than the commercial pecan shell char. Synthesized pecan shell char adsorbed the least NDMA. The initial mass transfer rate was controlled by film diffusion for all the tested carbons. Intraparticle diffusion was negligible for CSC but controlled a noticeable segment of adsorption for the other carbons. Column breakthrough trends did not follow expected patterns: breakthrough occurred much more rapidly than models indicated, and faster flowrates resulted in slower breakthrough and more apparent adsorption.

619077 Characterization of Cooling Water Blowdown in District Cooling and Determination of Its Effect on Wastewater Treatment Plants

Abstract:

The District Cooling (DC) process requires high volumes of water to operate cooling towers (CTs). In the State of Qatar, the use of Treated Sewage Effluent (TSE) in DC processes is mandated by the government for the purpose of sustainable management of limited water resources, efficient use of resources and to reduce Greenhouse Gas (GHG) emissions. The operation of CTs generates cooling water blowdown (CWBD), which is a contaminated liquid with very high concentrations of solids and chemicals. To adopt appropriate treatment and disposal strategies of CWBD, a full characterization of this wastewater is required. Th quality of CWBD depends on many parameters, including the quality of feed and make-up streams . Hence, the work presents a full characterization and analysis of the CWBD and TSE streams, including solids concentration, hardness, alkalinity, COD, pH, EC and toxicity. In addition, a bench scale study has been used to assess impacts of CWBD constituents on TSE and sludge. The results of the analysis and characterization work can be used in future research to simulate and validate the effect of CWBD constituents on the full-scale wastewater treatment process and, ultimately estimate the permitted volumes of CWBD that can be discharged to the WWTP without affecting the treatment process performance.

619079 Technical Review: Suitable Conventional and Emerging Water Treatment Systems for Cwbd Management

Abstract:

Cooling water blowdown (CWBD) is wastewater effluent from cooling towers that are used in industry for various purposes including cooling systems[1]–[3]. In the State of Qatar, district cooling facilities are used to supply large-scale cooling needs in high density developments such as business districts, commercial communities, university campuses, etc.Water quality of CWBD changes and depends on the quality of make-up water, the number of cycles of concentration (COC) and chemical additives used for reducing scaling and corrosion [4]. High concentrations of these contaminants in CWBD streams deem them unsuitable for discharge into surface water and wastewater treatment plants; and in some cases, may require pre-treatment. Therefore, the discharge regulations of the CWBD become more stringent to control the amount discharged to the eco-system, reduce makeup water requirements and to recycle and reuse as much water as possible. Hence, in the presented paper we investigate the existing and emerging water treatment technologies for the treatment of CWBD. We present the contaminants in CWBD and their corresponding impacts on the ecosystem; and we compare membrane and nonmembrane based conventional and emerging treatment technologies based on performance, cost, and limitations along with other factors. The membrane-based technologies reviewed in this article include membrane distillation (MD), reverse osmosis (RO), nanofiltration (NF), and vibratory shear enhanced membrane process (VSEP). For the nonmembrane-based technologies, electrocoagulation (EC), ballasted sand flocculation (BSF), and electrodialysis (ED) processes will be also discussed. EC and VSEP processes are emerging technologies with high performance and can be considered as the most suitable technologies for the treatment of CWBD compared to other processes. However, they are energy intensive and that can be controlled by utilizing renewable sources of energy. The study also compares international and regional (Gulf states) policies and regulations related to the standards for water contaminant discharge levels into waterbodies including surface water. Such regulations will help in delineating the limits of contaminants concentrations, consequently determining the most suitable technology for the treatment of these contaminants. The paper findings will guide the development of suitable policies for the management and discharge of CWBD into surface water and wastewater treatment plants.

References

[1] H. I. Abdel-shafy, M. A. Shoeib, M. A. El-khateeb, A. O. Youssef, and O. M. Hafez, "Electrochemical treatment of industrial cooling tower blowdown water using magnesium-rod electrode," *Water Resour. Ind.*, vol. 23, p. 100121, 2020.

[2] O. M. Hafez, M. A. Shoeib, M. A. El-Khateeb, H. I. Abdel-Shafy, and A. O. Youssef, "Removal of scale forming species from cooling tower blowdown water by electrocoagulation using different electrodes," *Chem. Eng. Res. Des.*, vol. 136, pp. 347–357, 2018.

[3] E. A. Anwer and B. A. Abdul Majeed, "Different Electrodes Connections in Electrocoagulation of Synthetic Blow down Water of Cooling Tower," *Iraqi J. Chem. Pet. Eng.*, vol. 21, no. 1, pp. 1–7, 2020.

[4] J. Ahmed, Y. Jamal, and M. Shujaatullah, "Recovery of cooling tower blowdown water through reverse osmosis (RO): review of water parameters affecting membrane fouling and pretreatment

schemes," *Desalin. Water Treat.*, vol. 189, no. June, pp. 9–17, 2020.

616729 CO_2/Brine Interaction with Reservoir Rock Under CO_2 Sequestration Conditions

Abstract:

The atmospheric CO_2 as well as other greenhouse gases have been considered as the main contributors to global climate change. To alleviate the effect of anthropogenic CO_2 on global climate change, many strategies are under development that can potentially remove CO_2 from the atmosphere. Among, many strategies, carbon capture and storage (CCS) is identified as the key strategy for reducing the atmospheric concentration of CO_2 from power plants and energy-intensive industries. In particular, CCS technologies could eliminate CO_2 emissions from power plants that use fossil fuels by separating CO_2 from plant flue gas and purifying, compressing, and transporting it *via* pipeline to underground geologic formations for permanent storage. The potential options for CO_2 storage include underground geological formations such as saline aquifers, depleted oil and gas reservoirs, unminable coal seams, hydrate storage, and CO_2 within enhanced geothermal systems. CO_2 storage in saline aquifers is recognized as one of the highest estimated storage capacities within underground geological formation. The Mount Simon formation in the Midwest region of the U.S. is considered as a potential candidate host reservoir for carbon storage. The Mount Simon formation is a deep saline aquifer and is a primary reservoir target for large scale carbon dioxide injection tests due to its proximity to CO_2 sources, large CO_2 storage capacity, favorable depth, thickness, permeability, porosity, and the presence of the thick overlying seals such as, Eau Claire formation. The Midwest Geological Sequestration Consortium (MGSC) has selected the Mount Simon formation as the reservoir at its major demonstration project. Obtain the knowledge of possible geochemically-induced changes to the permeability and porosity of host CO_2 storage sandstone will enable us to gain a deeper insight of the long-term reservoir behavior under the CO_2 storage conditions.

An experimental study of the interaction of CO_2/brine/rock on saline formations in a static system under CO_2 storage conditions was conducted. Chemical interactions in the Mount Simon sandstone exposure to CO_2 and brine under sequestration conditions were studied. Samples were exposed to the simulated *in-situ* reaction conditions for six months. Two core samples of Mount Simon sandstone (one is parallel to the bedding plane and the other is perpendicular to the bedding plane) were used. An Illinois Basin model brine at the temperature of 85 °C and CO_2 pressure of 23.8 MPa (3,500 psig) were applied to the reactors. CT, XRD, SEM, and brine, porosity, and permeability analyses were conducted prior and after the experiments.

The permeability and porosity measurements obtained from the sandstone sample showed a decrease after core was exposed to CO_2-saturated brine for six months. In addition, the permeability obtained from the core orientated parallel to the bedding plane is much larger than that measured from the core orientated perpendicular to the bedding plane. The combination of mineral dissolution and mineral precipitation occurring in the sample pores and cracks with the net effect of blocking of flow resulting in the observed the decrease in permeability. This observation suggests that mineral dissolution and mineral precipitation could occur in the host deposit altering its characteristics for CO_2 storage over time. In addition, the observed significantly difference in permeability obtained from the different orientation of core implies that during the CO_2 storage stage the CO_2 is more likely to flow toward the parallel than the perpendicular direction.

619043 Technical and Economic Assessment on Different Separation Technologies for Nitrogen Removal in Post Treatment of CO2 from Natural Gas

Abstract:

This paper presents the techno-economic analysis of separation technologies to remove high nitrogen concentration with a high concentration of CO_2 (~70%) in the natural gas stream. This is a very challenging process since the contaminants gasses are higher than the hydrocarbon gas stream. This study focuses on nitrogen removal after the bulk removal of the CO_2. The CO_2 stream's removal down to 20% leads to the high nitrogen content of ~10 mole % in the sales gas product. Hence, this study evaluated technologies that can remove nitrogen down to 1 mole% with minimal hydrocarbon loss to meet the LNG product specification. Furthermore, the hydrocarbon gas's high nitrogen concentration might lead to a lower heating value and corrosion issues in the downstream process equipment.

The technical and economic analysis was conducted to provide a semi-quantitative assessment to identify the potential gaps and find opportunities derived from the selected technology to meet the LNG product's nitrogen specification. The technologies were assessed based on four main criteria: technical, economic, operational, and strategic (technology advantage and risk) performance, including operating conditions, capital and operating costs, and technology readiness level. From the study, Cryogenic technology scored 76.3% of the overall rating for the nitrogen removal down to less than one mole% with the lowest cost other types of NRU technologies evaluated for nitrogen removal in the presence of high CO_2 (20 mol%). It is due to the higher pre-treatment cost of non-cryogenic technologies prior liquefaction process. Finally, the assessment leads to potential development, and process intensification of conventional cryogenic technology was identified for the nitrogen removal process and its' application.

AUTHOR INDEX